SURVEYING MATHEMATICS MADE SIMPLE

An original Book by

Jim Crume P.L.S., M.S., CFedS

Co-Authors
Cindy Crume
Bridget Crume
Troy Ray R.L.S.
Mark Sandwick P.L.S.

KINDLE and PRINTED EDITIONS

*****

PUBLISHED BY:

Jim Crume P.L.S., M.S., CFeds

Construction Staking

First publication: May, 2014

Cover photo Light Rail Construction
Mesa, Arizona 2014

TERMS AND CONDITIONS

The content of the pages of this book is for your general information and use only. It is subject to change without notice.

Neither we nor any third parties provide any warranty or guarantee as to the accuracy, timeliness, performance, completeness or suitability of the information and materials found or offered in this book for any particular purpose. You acknowledge that such information and materials may contain inaccuracies or errors and we expressly exclude liability for any such inaccuracies or errors to the fullest extent permitted by law.

Your use of any information or materials in this book is entirely at your own risk, for which we shall not be liable. It shall be your own responsibility to ensure that any products, services or information available in this book meet your specific requirements.

This book is covered by the Kindle Direct Publishing and/or CreateSpace Terms and Conditions.

This book may not be further reproduced or circulated in any form, including email. Any reproduction or editing by any means mechanical or electronic without the explicit written permission of Jim Crume is expressly prohibited.

TABLE OF CONTENTS

4

INTRODUCTION

Straight forward Step-by-Step instructions.

This book is just one part in a series of digital and printed books on Surveying Mathematics Made Simple. The subject matter in this book will utilize the methods and formulas that are covered in the books that precede it. If you have not read the preceding books, you are encouraged to review a copy before proceeding forward with this book.

For a list of books in this series, please visit:

http://www.cc4w.net/ebooks.html

Prerequisites for this book:

A basic knowledge of geometry, algebra and trigonometry is required for the explanations shown in this book.

Definition: Construction Staking - wikipedia: (a.k.a. "lay-out" or "setting-out") is to stake out reference points and markers that will guide the construction of new structures such as roads, bridges, buildings, subdivisions, etc. These markers are usually staked out according to a suitable coordinate system selected for the project.

5

EQUIPMENT AND MATERIALS

The equipment and materials needed for construction staking will vary depending upon the type of staking that is required.

The following are the survey equipment that will be needed: Survey grade GPS, Data Collector, Total Station, Robotic Total Station, Prism Pole, Level, Digital Level, Level Rods.

The following materials and hand tools will be needed: Sledge hammer (small and large), nails, lath (2' to 4'), hubs (various sizes and lengths), flagging (several colors), paint (several colors), rebar and caps for property corners (if in a subdivision).

A survey grade GPS is suitable for staking out horizontal positions and establishing horizontal control. For vertical positions, the GPS is suitable for vertical precisions that are equal to or larger than 0.10' such as slope stakes and rough grades.

A total station is suitable for staking out horizontal and vertical positions for precisions of 0.01' to 0.02' depending on the model.

A conventional level is suitable for running elevations at precisions of 0.01'.

For tighter vertical precisions, a digital level will be needed.

Use the following table to determine which equipment will be needed for your staking precision requirements.

Measurement Precisions		
	Horizontal	Vertical
GPS	0.02'	0.10' +/-
Total Station	0.02'	0.02'
Direct Level	------	0.01'
Digital Level	------	0.005'

The methods required to achieve the precisions needed will vary depending upon the equipment. GPS technologies continue to improve along with the horizontal and vertical precisions. Refer to the manufacturers specifications to insure that the equipment being used meets or exceeds the precisions needed.

HORIZONTAL AND VERTICAL CONTROL

The first task that needs to be completed prior to any staking is to verify the horizontal and vertical control. The design plans will indicate what horizontal and vertical control has been established for the construction project. This information is usually located on the cover sheet or geometric control sheet. On some design projects a local benchmark will be noted on the plans that is near the construction site. Other items that need to be reviewed before staking are the details that are shown for building pads, water, sewer, typical cross-sections and so forth that contain controlling design grades.

Review the design plans to look for the horizontal and vertical control. Look for saw cut lines where the new construction features tie into existing pavement, curb and gutter or any other design constraints that need to be match into.

Run level loops from the primary vertical control benchmark to all local benchmarks and design constraints to verify that the vertical elevations shown on the design plans are shown accurately. It is very important that these elevations be checked to insure that the staking will match the design elevations.

DO NOT assume that these vertical checks have been made by the design engineer and that you do not need to check them. You will save yourself a lot of headache by performing these checks in advance of any pre-calculations or staking.

Any differences between the level loop and the design plans need to be brought to the attention of the design engineer before doing any staking.

You will need to verify the horizontal control to insure that the design plans are tied to some controlling line such as a road centerline, property line, building, coordinates or other horizontal ties that relate the design plans to some physical controlling feature on the ground. You may need to coordinate with the design engineer to determine what the controlling lines are if they are not clearly shown on the design plans.

Once you have verified the primary horizontal and vertical control, you may want to set secondary control points throughout the project site in areas that will not be disturbed during construction. You will want to make sure that you have line of sight between primary and secondary control points that can be utilized as back sights for non GPS equipment such as Total Stations.

You can never have too many secondary control points. Place plenty of guard lath around your control points and paint them so that they are very visible.

In your scope of work, you should have a clause that any control points that are destroyed by the contractor will be replaced at additional costs. Nothing is more frustrating than getting to the job site to start your day and your control points are missing.

It is also a good idea to write the Point number, Northing, Easting and Elevation on the lath for each of the control points. Mark the bottom of the lath with (LVL) for level ran elevations or (GPS) for GPS ran elevations.

You will also want to document your control information in a drawing showing the meta data for each point and it's relationship with the construction site. You will want to share this information with the contractor so that they are utilizing the same information you are.

Many construction companies are implementing GPS controlled grading equipment which requires survey control information be synchronized with their GPS equipment.

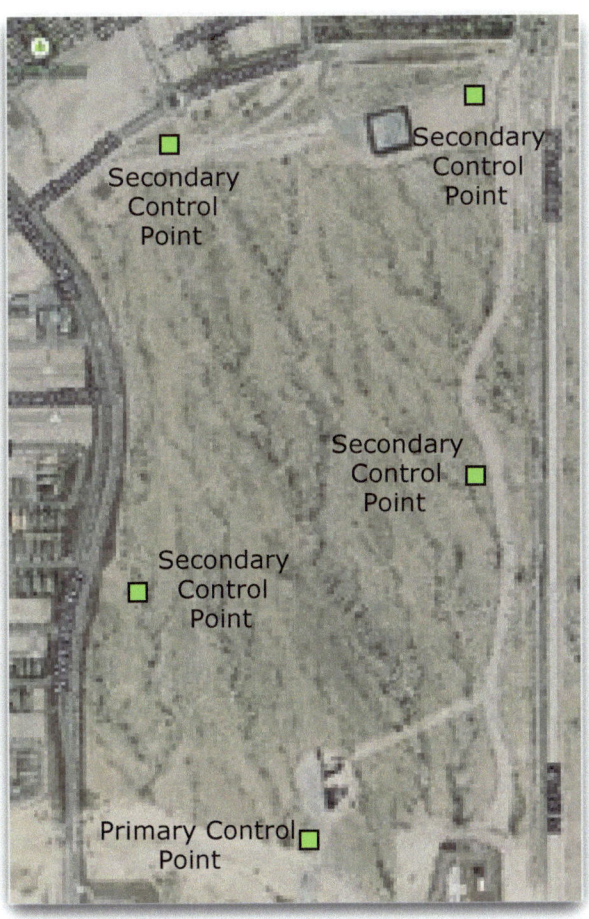

Typical Control Layout

11

GRID VS GROUND COORDINATES VS CALIBRATION/LOCALIZATION

Design plans are almost always drawn with ground distances. The stakes that are set on the ground need to represent the same datum as the design plans, which are usually ground coordinates and ground distances.

GPS utilizes geographical coordinates. The trick when using GPS is to relate ground coordinates to geographical coordinates.

This can be accomplished in two ways. One method is to convert ground coordinates to grid coordinates for the State Plane Zone that you are working in and then let the software in the data collector perform the conversion of grid coordinates to geographical coordinates.

A second method is to perform a least squares adjustment (a.k.a. Calibration/Localization) to known control points with ground coordinates at the project site.

Both methods have their pros and cons.

If you don't have a lot of experience working with state plane coordinates, then method one will be foreign to you and difficult to understand.

Most surveyors opt for method two, Calibration/Localization to known ground coordinates which can also be difficult to understand if you don't have any experience with least squares adjustment.

Calibration/Localization can be your friend and at the same time your worst enemy if you do not understand how calibration/ localization works.

If you utilize a one point calibration/localization to a control point then the geographical coordinates in the GPS are translated in the Northing, Easting and Elevation to this one point. All bearings will be grid bearings and distances will be based upon the method used to create the coordinates which will probably be ground distances from the design plans.

If a two point calibration/localization is utilized with two known control points, then a rotational and scaling component will be introduced with the least squares adjustment. This is where it gets tricky. Most data collector software have settings to use the rotational component only with no scaling. If scaling is introduced then all of the coordinates that were calculated and entered into the data collector will be rotated and scaled. This may or may not be a good thing.

If a multiple point calibration/localization is applied, then you will have a rotational and scaling being applied to all of the coordinates for the project that will be different than a one or two point calibration/localization.

It is imperative that if calibration/localization is utilized, that the same control points be utilized for each GPS setup. If one of the control points that was initially used gets destroyed, then you will not be able to perform the same least squares adjustment that was used the first time and you will have a different rotational and scaling component.

The vertical component of a Least Squares Adjustment is not widely understood but can be described as follows: For a one point calibration/localization, all coordinate points are adjusted to only one point and the vertical plane stays horizontal. If two or more control points are utilized for a vertical calibration/localization, then you will introduce an tilted inclined vertical plane that is no longer horizontal. This can become a real issue if any one of the control points becomes lost or destroyed or the elevation changes due to being damaged.

I have utilized both methods described above. My personal preference is to work in State Plane Coordinates. If I need to perform a calibration/localization, I generally pick one control point for the primary control and perform a one point calibration/localization for horizontal and vertical adjustments.

Keep in mind that a multi-point calibration/localization has a useful purpose if used properly and on the right type of survey. An understanding of the least squares adjustment will go a long way into assisting you on when to and when not to utilize this method.

For large construction projects, it is a good idea to separate it into zones to minimize the effect of working on the earths surface which is a sphere. I have worked on large construction projects that have spanned over 20 miles and the horizontal and vertical control needs to be looked at very closely. Separating the project into control zones worked out very well.

DESIGN PLANS TO USABLE COORDINATES

Before any stakes can be set for construction purposes, the design plans must be converted to Point Number, Northing, Eastings, Elevations and Description. This can be accomplished by utilizing the design electronic CADD files, if they are available from the design engineer, and officially approved design plans. NOTE: Only use officially approved design plans. On some projects the electronic files are not available or won't be released by the engineer.

For small projects, the calculations can be done in the field on the fly. For most construction projects, the calculations should be done in the office under a controlled environment to minimize mistakes. I have personally calculated on the tailgate of my truck in the field and I would much rather do the calculations in the office where checks can be made prior to staking.

Coordinates (Pt, N, E, EL, Desc format) will need to be created for each controlling grade point and/or angle point for the construction feature that the contractor will need to build the project. You will need to consult with the construction supervisor to determine exactly which points are going to be needed for construction and what type of offset will be needed. This will vary between construction companies and the type of construction. For example, curb and gutter will require a certain amount of stakes versus a retention basin.

A point will need to be calculated for each design feature at its finish grade position. Offsets can be calculated in the field with a competent crew chief and the right data collector.

References:

There are several books and mobile apps available in this Math-Series that will aide in developing coordinates that can be utilized with the GPS.

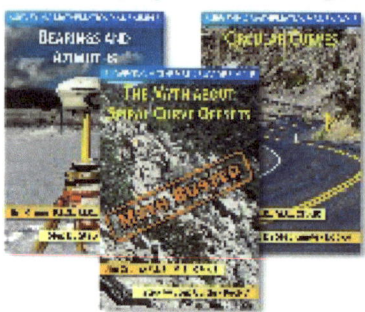

SURVEY FIELD PACKETS

You will need to prepare field packets for the survey crews giving them instructions on what needs to be staked for each day according to the schedule provided by the contractor.

The field packet contains drawings with point numbers of the points that need to be staked and a Master ASCII file that is needed which contains all of the points with coordinate values for each day of staking. It also contains the Horizontal and Vertical Control that needs to be utilized. On large staking projects it is a good idea to assign point ranges for different features. (i.e. Sewer Pt's 1000 to 2000, etc.)

This information can be delivered in person or through cloud based technology such as Dropbox or some other suitable method.

Time is money and any method that can be implemented to keep the survey crews in the field working is better than having them commute back and forth to the office to get what they need for each day.

The use of a Tablet with Internet connections can be beneficial for wirelessly transferring data between the office and the field.

TYPICAL ORDER OF STAKING PRECEDENCE

- Prepare Field Packets (1)

- Clear and Grubbing

- Rough Grade (2)

- Saw Cut Lines

- Sewer

- Water

- Dry Utilities

- Sleeves

- Curb

- Subgrade

- ABC Bluetops

- As-Built - Field

- As-Built - Office

(1) The field packets maps should be supplied to the contractor at the mass grade stage to aide their grade checker in following the stakes that are being set.

(2) The surveyor should coordinate with the contractor to utilize designed roadways as haul roads. This will help minimize the rough grade stakes from being obliterated.

CLEARING AND GRUBBING

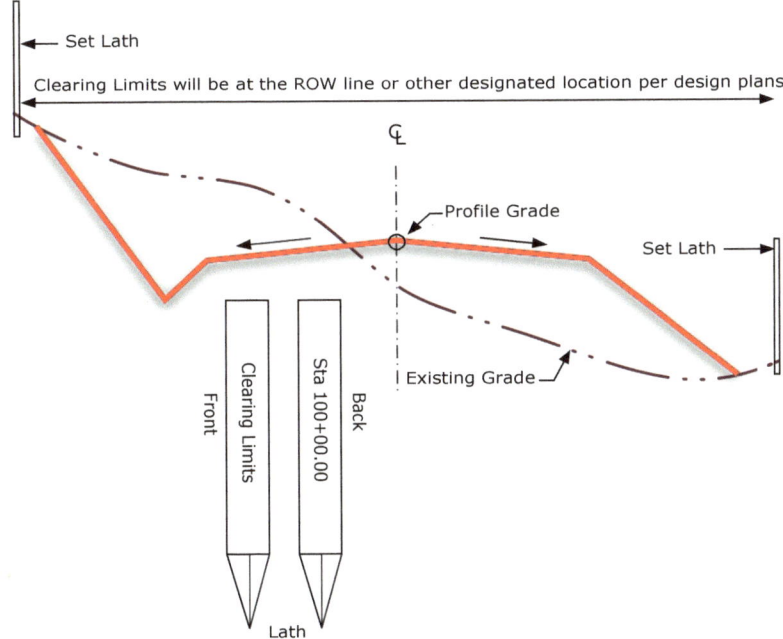

Field Procedures

One of the first stakes that you will need to set is for clearing and grubbing. The site will need to be cleared of any existing debris before rough grading.

The design plans will indicate where the outer limits are for construction. The contractor will need lath set for this perimeter so that all construction activities are contained inside of this perimeter.

These markings usually consist of a nail with flagging and a 4' lath that is flagged or painted with florescent colors. The lath will be marked with "Clearing Limits" or some designation as defined by the contractor.

These stakes are usually set to plus or minus 0.10'.

ROUGH GRADE STAKES

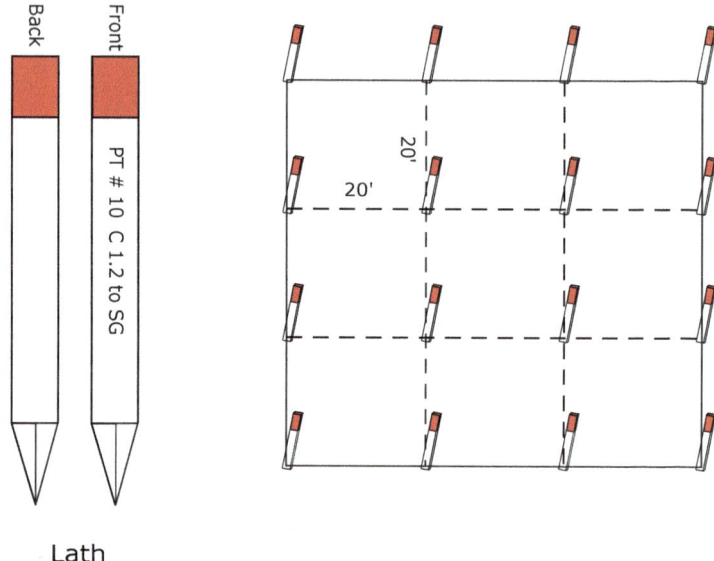

Lath

Typical Grid Pattern

Field Procedures

Each construction site is different. Some sites will require rough grade stakes that will aide the contractor in removing or depositing material to get the site close to finish sub-grade.

These stakes are usually set to plus or minus 0.10'.

The location of these stakes will be controlled by the design features on the plans such as building pads, roads, streets, walls, etc.

A grid pattern may be needed for large sites. If the contractor is using GPS controlled grading equipment, fewer stakes may be needed. If rough grade for site, cuts or fills, are more than 5' then the surveyor and contractor need to discuss getting the site closer to grade prior to staking.

SLOPE STAKES

Of all of the types of construction staking, Slope Stakes are the most challenging. The reason being is that the exact position of the slope stake is not known until you make some measurements both horizontally and vertically in the field. The position is then adjusted as needed and new horizontal and vertical measurements are taken. The correct position (catch point) can only be determined after several adjustments are made until the true intersection of the design grade slope and existing grade is found.

Slope staking with a GPS can be challenging as well. The manufacture and software for each data collector type will require different settings that will need to be completed to find the catch point. You will need to consult the user manual for directions on slope staking with a GPS.

To better understand the process used to determine the catch point, this book will described the step by step process needed using conventional equipment (chain, level and level rod) that will need to be adapted to your GPS equipment.

Let's look at a basic example shown below.

SLOPE STAKING

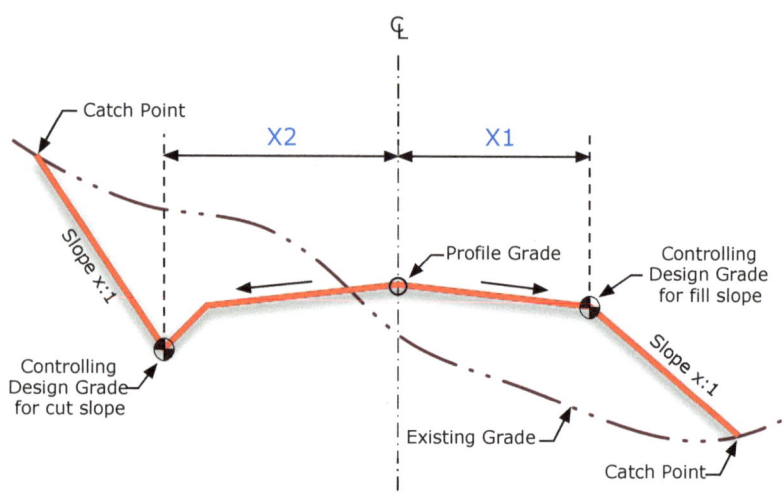

Typical Roadway Section

This example is of a typical roadway section however the principles shown herein will apply to any design feature where a design grade slope intersects the existing grade, such as a new ditch, retention pond, detention pond, etc.

For this roadway section there is a cut slope and a fill slope. The basic known points that are needed are the controlling profile grade, controlling design grade (where the design slope begins) and the new design slope (i.e. 3:1, 4:1, 6:1, etc.). This information will be shown on the design plans.

Lets look at the components for the Fill slope first.

The image below shows a typical fill slope.

SLOPE STAKING

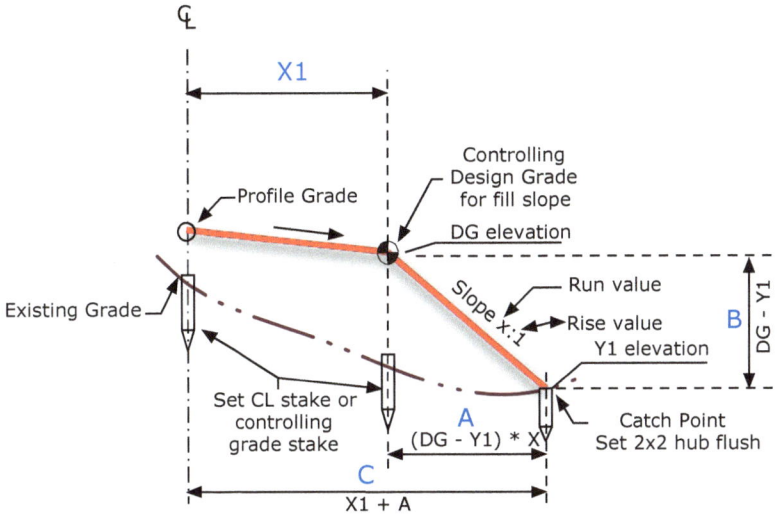

Note: Catch point found when A = B * Run value

Typical Fill Slope

Field Procedures

1. Set a stake at the controlling profile grade and/or the controlling design grade.

2. Determine the DG elevation from the design plans.

3. Use a GPS or direct level to measure the ground elevation at the DG location.

4. Determine the Fill value at the DG location (DG elevation - ground elevation).

5. Multiply the Fill value by the slope run value to get distance (A).

6. As a starting point, measure this distance out (A) from the controlling design grade stake and now measure the elevation (Y1) at this location.

7. Determine the Fill value (B) by subtracting the Y1 elevation from the DG elevation.

8. Multiply the Fill value by the slope run value to get a new distance (A).

9. If the distance measured from the controlling grade stake equals the new distance (A), then the catch point has been found. Set a hub flush with the ground and mark the lath as shown at the end of this section.

10. If not, measure out the new distance (A) from the controlling grade stake and measure a new elevation (Y1) at this location.

11. Repeat steps 7 through 10 until the catch point is found.

The catch point will be found when A = B * Run value.

Example

DG elevation = 100.00
DG ground elevation = 96.00
Slope = 6:1 and X1 = 12.00

Fill = 100.00 - 96.00 = 4.00 (B)
Fill * Run = 4.00 * 6 = 24.00 (A)
Measure out 24.00 from DG

New ground elevation = 95.40 (Y1)
Fill = 100.00 - 95.40 = 4.60 (B)
Fill * Run = 4.60 * 6 = 27.60 (A)
Measure out an additional 27.60 - 24.00 = 3.60 feet.

New ground elevation = 95.40 (Y1)
(A) = (B) * 6, catch point found

(C) = 27.60 + 12.00 = 39.60 from centerline.
(B) = 4.60

(A) = 27.60

Lets now look at the components for the Cut slope.

The image below shows a typical Cut slope.

SLOPE STAKING

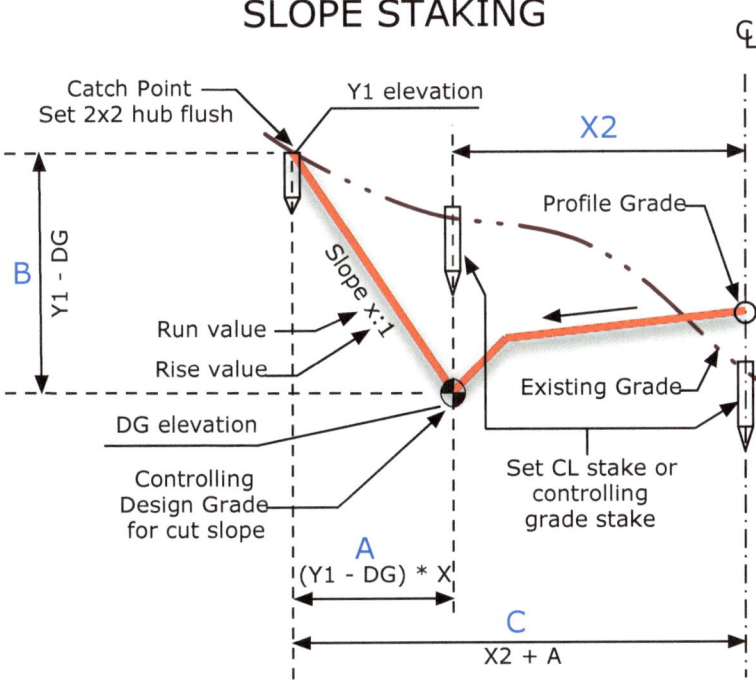

Note: Catch point found when A = B * Run value

Typical Cut Slope

Field Procedures

1. Set a stake at the controlling profile grade and/or the controlling design grade.

2. Determine the DG elevation from the design plans.

3. Use a GPS or direct level to measure the ground elevation at the DG location.

4. Determine the Cut value at the DG location (DG elevation - ground elevation).

5. Multiply the Cut value by the slope run value to get distance (A).

6. As a starting point, measure this distance out (A) from the controlling design grade stake and now measure the elevation (Y1) at this location.

7. Determine the Cut value (B) by subtracting the Y1 elevation from the DG elevation.

8. Multiply the Cut value by the slope run value to get a new distance (A).

9. If the distance measured from the controlling grade stake equals the new distance (A), then the catch point has been found. Set a hub flush with the ground and mark the lath as shown at the end of this section.

10. If not, measure out the new distance (A) from the controlling grade stake and measure a new elevation (Y1) at this location.

11. Repeat steps 7 through 10 until the catch point is found.

The catch point will be found when A = B * Run value.

Example

DG elevation = 98.00
DG ground elevation = 103.00
Slope = 3:1 and X2 = 14.00

Cut = 103.00 - 98.00 = 5.00 (B)
Cut * Run = 5.00 * 3 = 15.00 (A)
Measure out 15.00 from DG

New ground elevation = 108.00 (Y1)
Cut = 108.00 - 98.00 = 10.00 (B)
Cut * Run = 10.00 * 3 = 30.00 (A)
Measure out an additional 30.00 - 15.00 = 15.00 feet.

Keep moving out until catch point is found.

New ground elevation = 110.50 (Y1)
Cut = 110.50 - 98.00 = 12.50 (B) * 3 = 37.50 (A)
(A) = (B) * 3, catch point found

(C) = 37.50 + 14.00 = 51.50 from centerline.
(B) = 12.50
(A) = 37.50

MARKINGS ON THE SLOPE STAKES

The following exhibits show the markings on the slope stakes and what they represent.

SLOPE STAKING

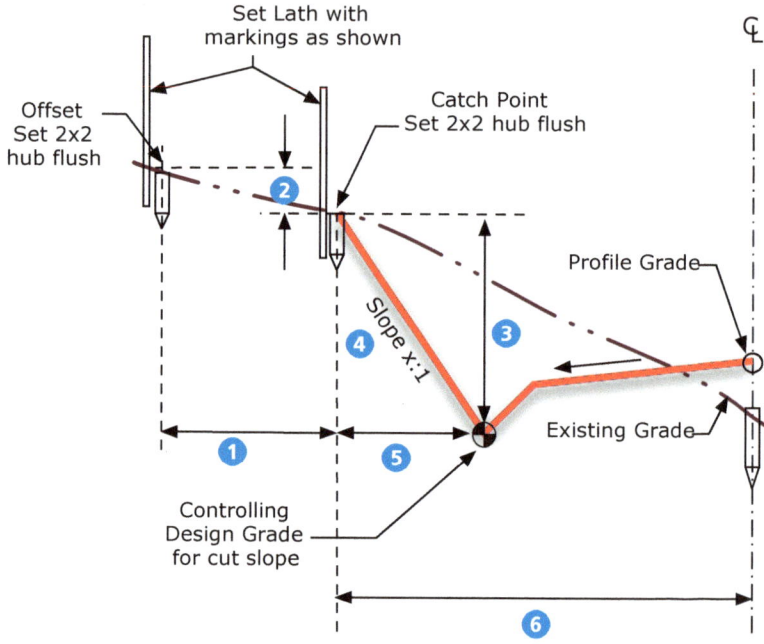

Typical marking on the Lath - Cut

SLOPE STAKING

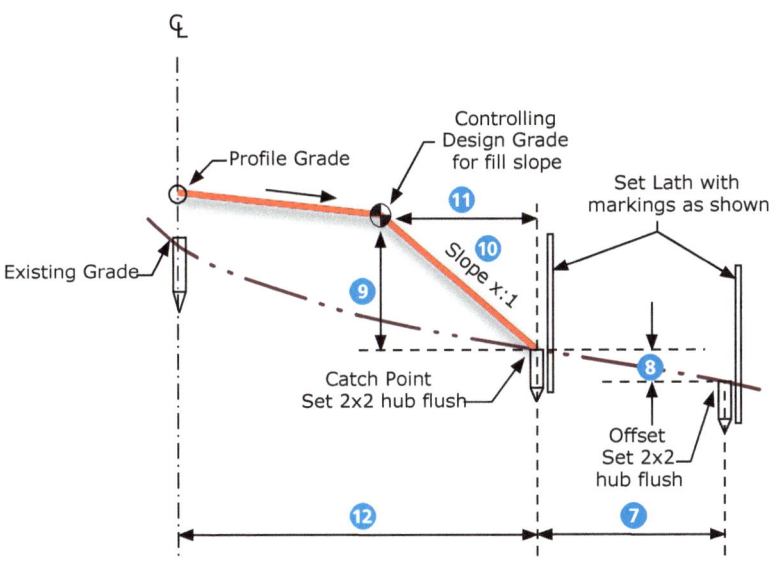

Typical marking on the Lath - Fill

SLOPE STAKING

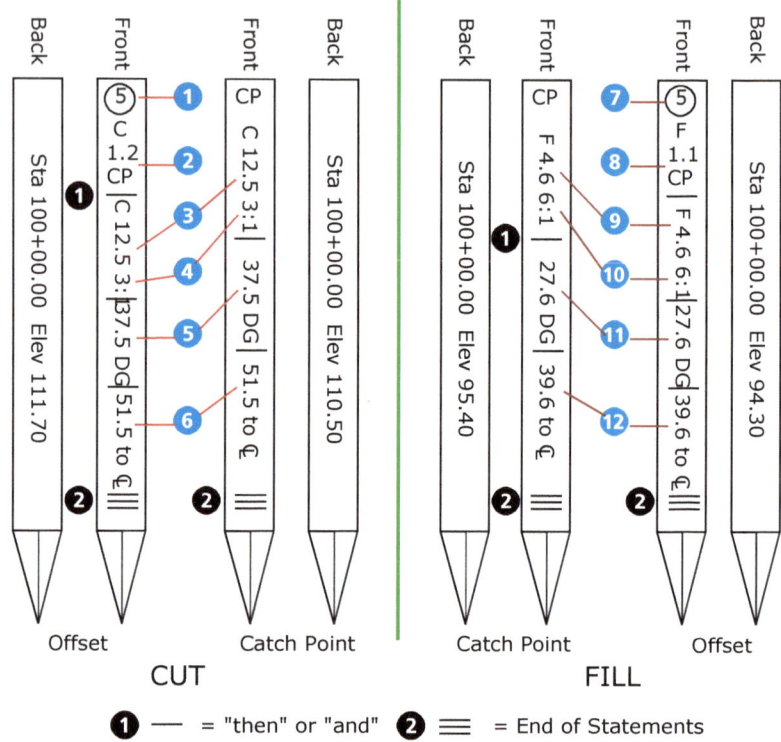

CUT

FILL

1 — = "then" or "and" **2** ≡ = End of Statements

Typical marking on the Lath

CENTERLINE STAKING

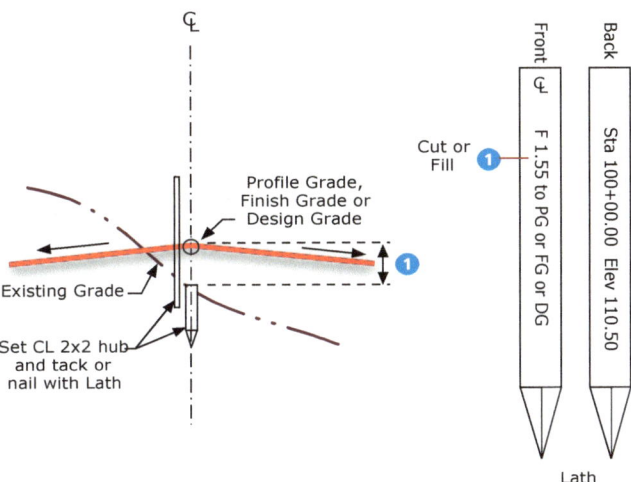

Controlling Centerline

Field Procedures

Centerline stakes (hub or nail) may be required along roads, highways, paths, pipelines, etc.

Set grade stakes and bluetops at intervals per the clients instructions (usually 25' on tangent lines and 25' on curves).

Run elevations to each of the stakes using the survey equipment that will meet the precision requirements needed. See the section titled "Equipment and Materials" described earlier in this book.

Subtract the stake elevation (SE) from the design grade (DG) to determine the cut or fill amount.

(DG - SE = Cut or Fill)

If curb has been set, then the centerline hubs can be based off the high side of curb to establish the crown of road.

Mark the lath as shown above.

PIPE STAKING

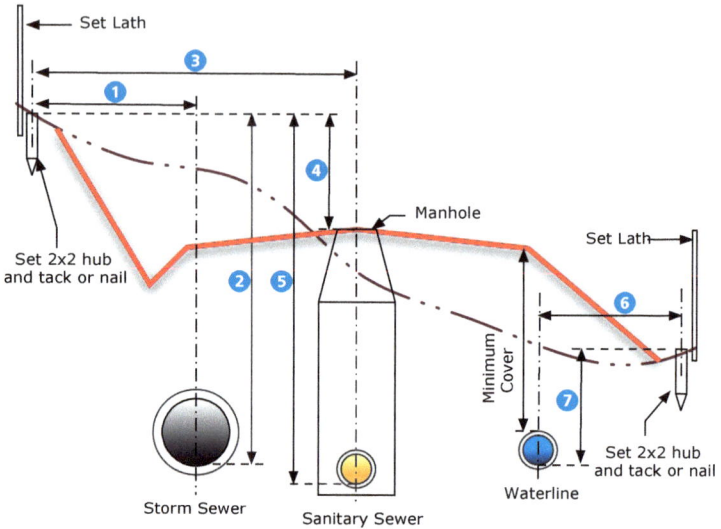

Storm Sewer, Sanitary Sewer and Waterlines

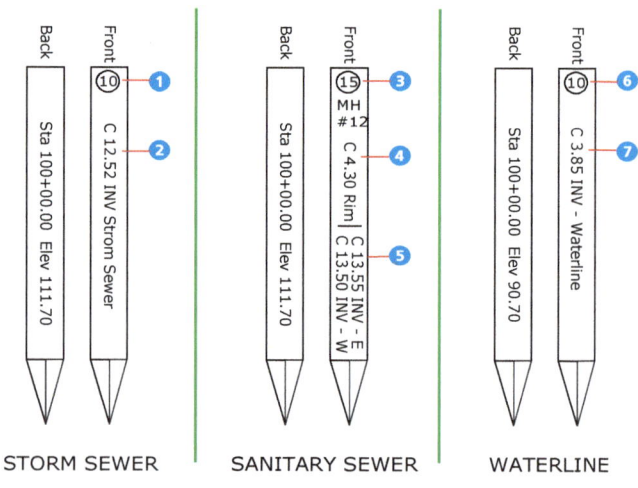

Typical marking on Lath

Field Procedures

Pipeline offset stakes (hub or nail) will be needed for any underground utility construction such as Storm/Sanitary Sewer Lines, Water lines, etc. that have controlling elevations and grades.

Set stakes at an offset distance that will not be disturbed during construction. Set offset stakes at all grade breaks, manholes, and so forth and at intervals between grade breaks per the clients instructions (normally at 25' intervals).

Run elevations to each of the offset stakes using the survey equipment that will meet the precision requirements needed. See the section titled "Equipment and Materials" described earlier in this book.

Subtract the offset stake elevation (SE) from the design grade (DG) to determine the cut amount (Normally the design grade will be the invert (flow line) elevations on Manholes and pipelines. In some cases it maybe to the top of pipe).

(DG - SE = Cut)

Mark the lath as shown above.

When staking Fire Hydrants (FH), be sure to stake the curb limits to confirm/prove to the contractor that the FH are being installed per the plan details. The same process should be used for the staking of sewer services.

FINISH GRADE - BLUETOPS

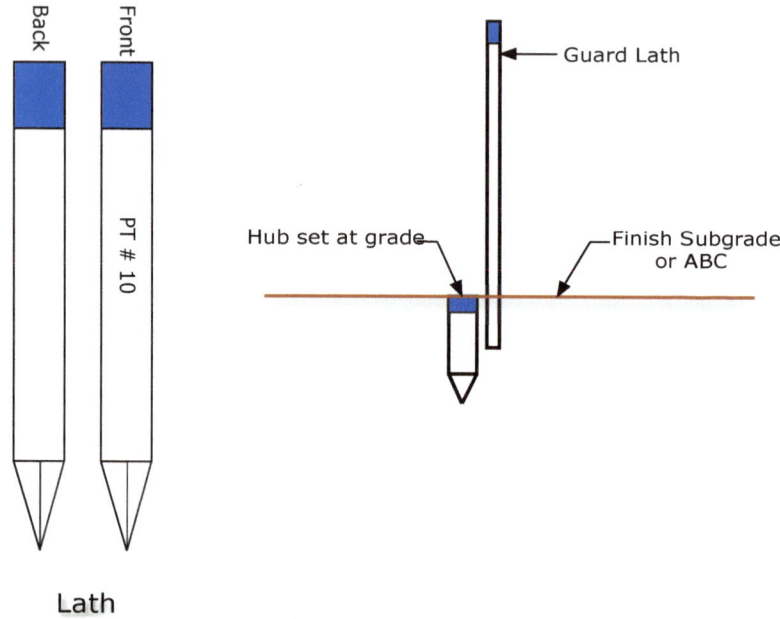

Typical Bluetop

Field Procedures

Finish grade stakes a.k.a. Bluetops (Hub) will be set to grade and painted blue.

These stakes are usually set to plus or minus 0.02'.

The location of these stakes will be controlled by the design features on the plans such as building pads, roads, streets, walls, etc.

A grid pattern may be needed for large sites. If the contractor is using GPS controlled grading equipment, fewer stakes may be needed.

DRY UTILITIES

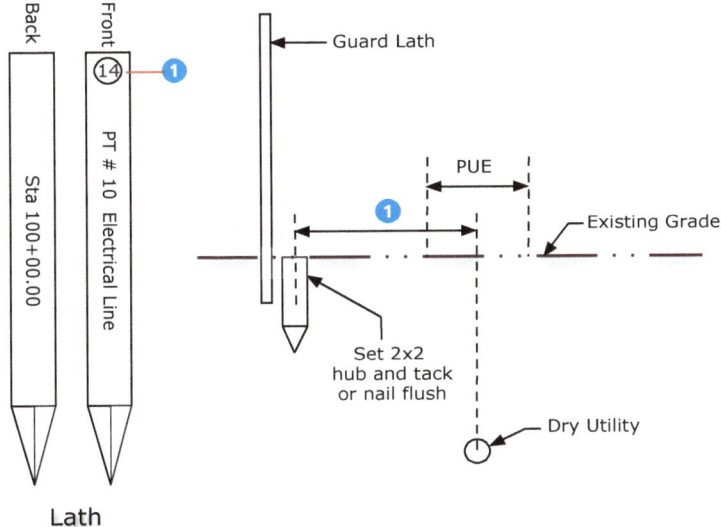

Electrical, Cable-TV, Fiber Optics, Telephone, etc

Field Procedures

Dry utilities require an offset stake (hub or nail) for location only. Grades may be required depending on the unfinished conditions of the site. Most of these stakes are related to the top of curb.

These offset stakes are set at an offset distance so that they do not get disturbed during construction. (Usually a 10' offset from back of PUE. i.e. 14' to centerline of 8' PUE)

The location and intervals of these offset stakes will be controlled by the design features on the plans such as building pads, roads, streets, walls, etc. Coordinate the setting of these offset stakes with the contractor.

Mark the lath as shown above.

CURB AND GUTTER

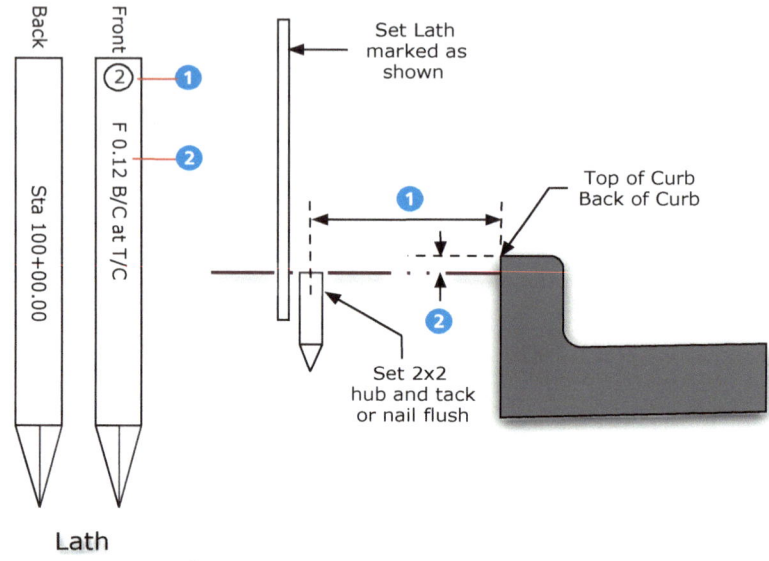

Typical Curb and Gutter

Field Procedures

Curb stakes (hub or nail) will be required for all new curb construction and all types of curbs, those with gutters and those without. The grade stake will be set and graded to the "Top of Curb" at the "Back of Curb".

These offset stakes are set at an offset distance so that they do not get disturbed during construction and are close enough for setting up the string line for curb machines. (Usually a 2' offset). The contractor will want to know where the catch basins, wings and returns are located so that the curb machine can bypass these areas.

The location and intervals of these stakes will be controlled by the design features on the plans such as curb returns and grade breaks. (Intervals of 25' or as directed by the contractor)

36

Run elevations to each of the offset stakes using the survey equipment that will meet the precision requirements needed. See the section titled "Equipment and Materials" described earlier in this book.

Subtract the offset stake elevation (SE) from the design grade (DG) to determine the cut or fill amount.

(DG - SE = Cut or Fill)

Mark the lath as shown above.

SIDEWALK

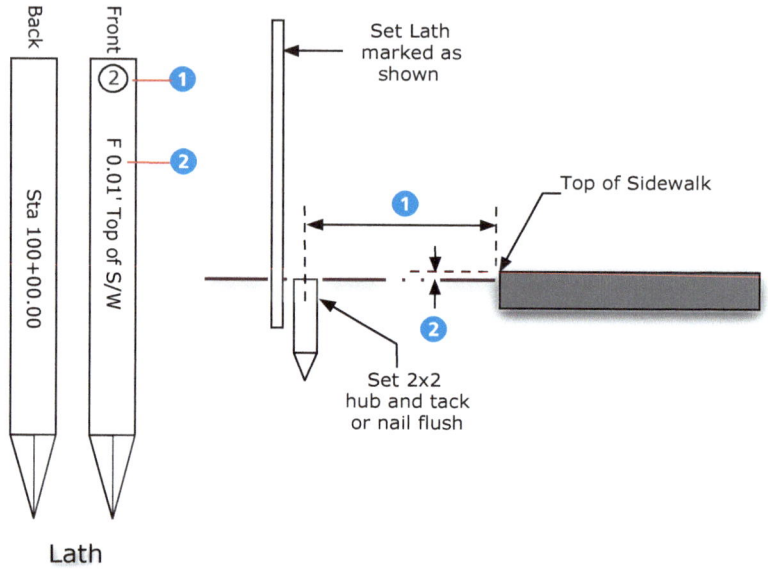

Typical Sidewalk

Field Procedures

Sidewalk stakes (hub or nail) will be required for all new sidewalk construction. The grade stake will be set and graded to the "Top of Sidewalk.

These offset stakes are set at an offset distance so that they do not get disturbed during construction and are close enough for setting up the string line for sidewalk machines. (Usually a 2' offset).

The location and intervals of these stakes will be controlled by the design features on the plans such as begin/end of curved segments and grade breaks. (Intervals of 25' or as directed by the contractor)

Run elevations to each of the offset stakes using the survey equipment that will meet the precision requirements needed.

See the section titled "Equipment and Materials" described earlier in this book.

Subtract the offset stake elevation (SE) from the design grade (DG) to determine the cut or fill amount.

(DG - SE = Cut or Fill)

Mark the lath as shown above.

DITCH STAKES

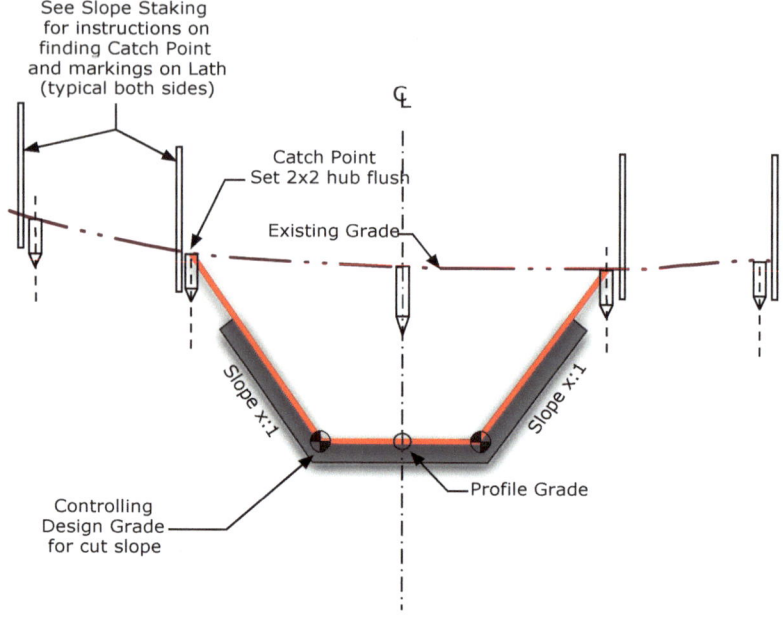

See Slope Staking for instructions on finding Catch Point and markings on Lath (typical both sides)

℄

Catch Point
Set 2x2 hub flush

Existing Grade

Slope X:1

Slope X:1

Controlling Design Grade for cut slope

Profile Grade

Typical concrete lined Ditch

Field Procedures

Slope stakes (Hub or nail) will be needed for all new ditch construction. See the section on "Slope Staking" in this book for the field procedures required to locate the catch point for the new slope.

The markings on the stake will vary to accommodate any changes in materials such as a concrete lined ditch. Coordinate with the contractor for any special markings that might be required.

40

SIGNS

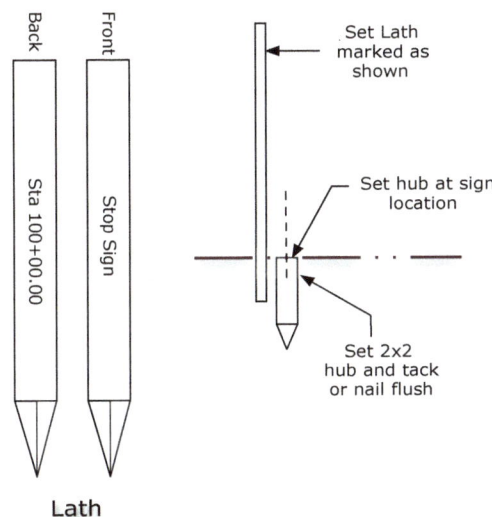

Set Lath marked as shown

Set hub at sign location

Set 2x2 hub and tack or nail flush

Back — Sta 100+00.00

Front — Stop Sign

Lath

Minor Signs

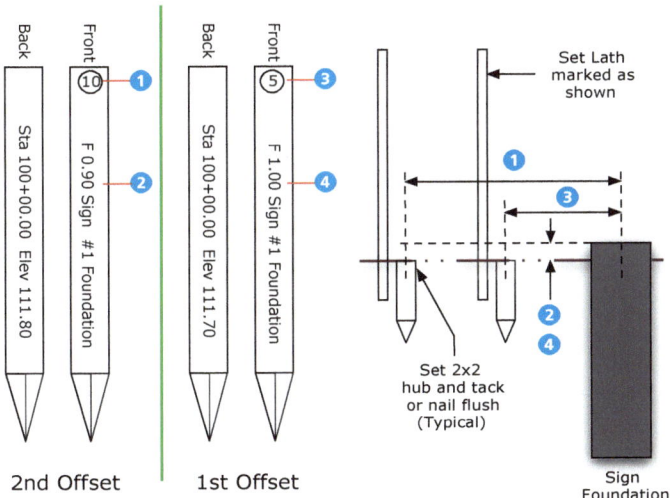

Set Lath marked as shown

Set 2x2 hub and tack or nail flush (Typical)

Sign Foundation

2nd Offset — Back: Sta 100+00.00 Elev 111.80 — Front: F 0.90 Sign #1 Foundation

1st Offset — Back: Sta 100+00.00 Elev 111.70 — Front: F 1.00 Sign #1 Foundation

Major Signs

41

Field Procedures

The type of signs that will need to be staked will vary on each construction project. They may range from simple one pole signs such as Stop, Yield, Caution, etc. to more complex signs such as elevated road signs on highways and freeways. The type of sign will determine the type of staking that will be needed.

Simple Signs

Set stakes (Hub or nail) at the center of the sign. Usually no grade is needed for one pole simple sign location.

Complex Signs

For signs that need to be graded and constructed in a precise location, two offset stakes (Hub or nail) will be required.

Run elevations to each of the offset stakes using the survey equipment that will meet the precision requirements needed. See the section titled "Equipment and Materials" described earlier in this book.

Subtract the offset stake elevation (SE) from the design grade (DG) to determine the cut or fill amount.

(DG - SE = Cut or Fill)

Mark the lath as shown above.

MINOR STRUCTURES

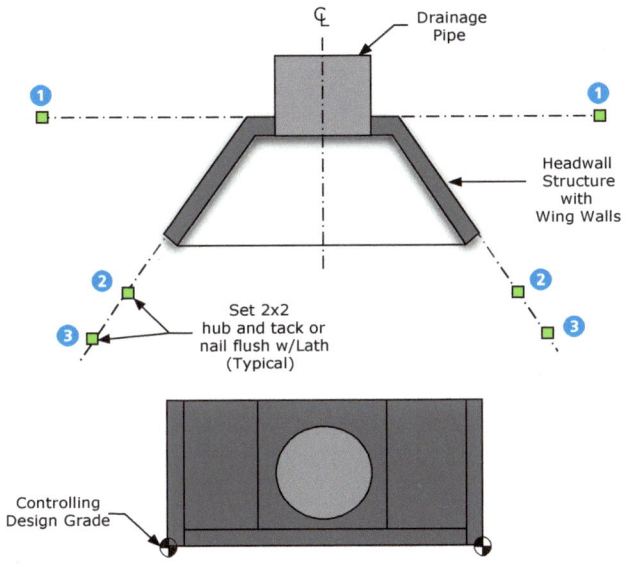

Typical Headwall Layout

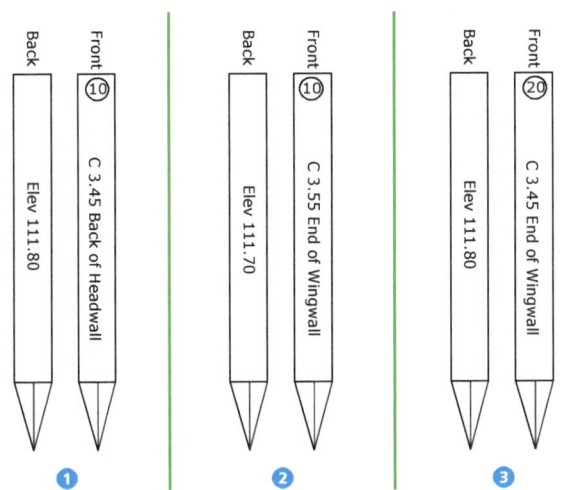

Typical markings on Lath

Field Procedures

Minor structures such as Culverts and Headwalls will require a combination of offset stakes (Hub or nail). The configuration of the design feature will help in determining the number of offset stakes that will be required. Always review the plan details before staking (field and office) to avoid blunders.

Coordinate with the contractor to determine the numbers of offset stakes that will be needed for them to build the minor structure.

Run elevations to each of the offset stakes using the survey equipment that will meet the precision requirements needed. See the section titled "Equipment and Materials" described earlier in this book.

Subtract the offset stake elevation (SE) from the design grade (DG) to determine the cut or fill amount.

(DG - SE = Cut or Fill)

Mark the lath as shown above.

MAJOR STRUCTURES

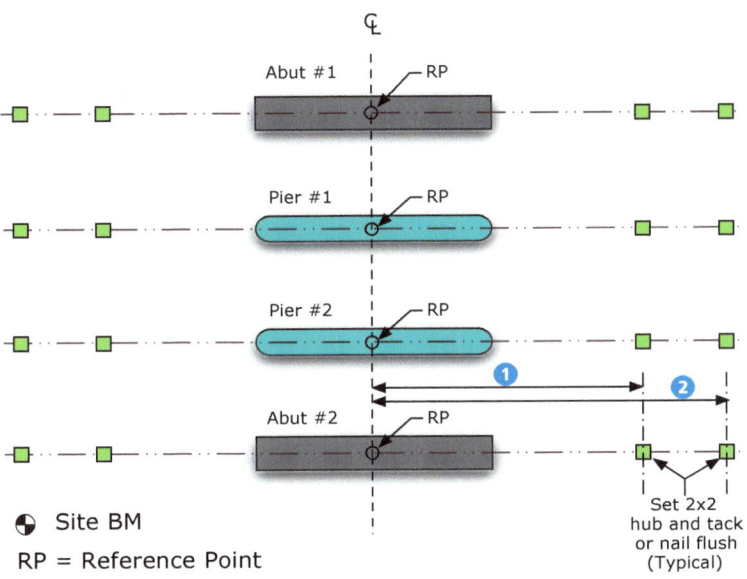

Site BM

RP = Reference Point

Set 2x2 hub and tack or nail flush (Typical)

Typical Bridge Layout

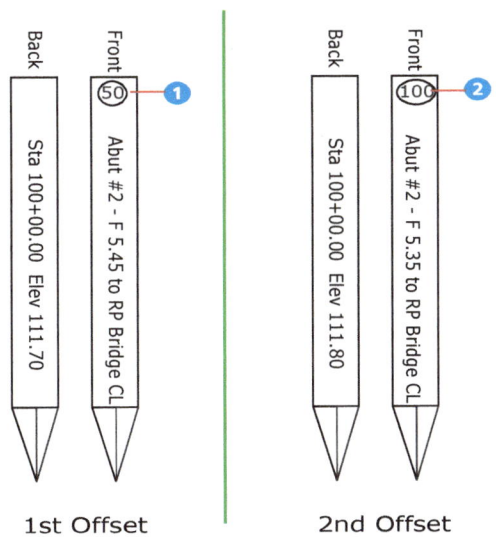

1st Offset

2nd Offset

Typical marking on Lath

45

Field Procedures

Major structures such as Bridges and Box Culverts will require multiple offset stakes (Hub or nail). The configuration of the design feature will help in determining the number of offset stakes that will be required.

Coordinate with the contractor to determine the numbers of offset stakes that will be needed for them to build the major structure.

A Bench Mark is normally located near the major structure in an area where it will not be disturbed during construction.

Several sets of offset stakes will be required for each of the abutments and piers that the contractor will utilize to run string lines to set their forms. Coordinate with the contractor to determine the number and type of stakes. They may utilize batter boards and will need stakes set to accommodate this method.

Run elevations to the benchmark and each of the offset stakes using the survey equipment that will meet the precision requirements needed. See the section titled "Equipment and Materials" described earlier in this book.

Subtract the offset stake elevation (SE) from the design grade (DG) to determine the cut or fill amount.

(DG - SE = Cut or Fill)

Mark the lath as shown above.

Additional grades may be required on the bridge deck for the Bidwell paving machine. Coordinate this staking with the contractor.

Bidwell Paving Machine

Typical Batter Board

WALL AND FENCE STAKES

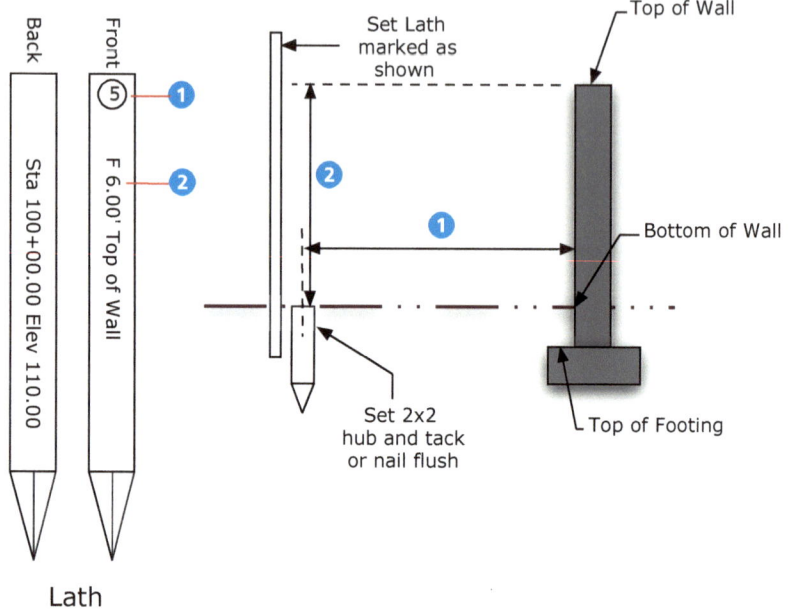

Typical Wall Staking

Field Procedures

Walls and Fences will require offset stakes (Hub or Nail). The configuration of the design feature will help in determining the number of offset stakes that will be required.

Coordinate with the contractor to determine the numbers of offset stakes that will be needed for them to build the wall or fence.

Run elevations to each of the offset stakes using the survey equipment that will meet the precision requirements needed. See the section titled "Equipment and Materials" described earlier in this book.

48

The design grade can be to the "Top of Wall", "Top of Footing" or "Bottom of Wall" which is where the finish grade butts up against the wall. Coordinate with the contractor on what to use for the controlling grade.

Subtract the offset stake elevation (SE) from the design grade (DG) to determine the fill amount.

(DG - SE = Fill)

Mark the lath as shown above.

RIGHT OF WAY AND CENTERLINE STAKES

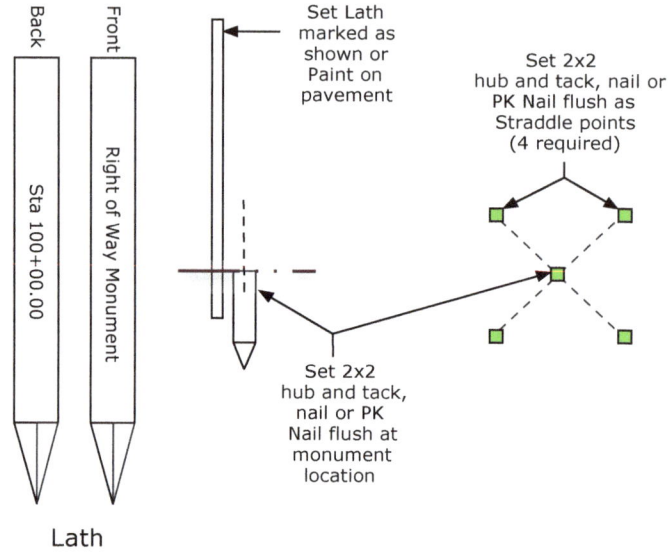

Typical Monument Staking

Field Procedures

Right of Way and Centerline monuments will need to be staked so that the contractor can set the final standard monument which is usually a cap in concrete or some other permanent type monument.

A Hub, nail or PK nail (when located in asphalt) will be set for the exact location of the monument. Two sets of straddle points on opposite sides of the true location will be set using a string line. Set the first straddle point about 1.5' offset then using a string from the first straddle point place it over the true location and set a second straddle point on the opposite side about 1.5'. Repeat this process for the second set of straddle points approximately perpendicular to the first set.

After the standard monument has been set, using a string line, connect first set of straddle points and make a mark on the cap. Connect the second set of straddle points and make a mark on the cap. Place a punch mark at the intersection of the two marks. Finish by stamping the cap with your registration number and any other markings that are required.

CONCLUSION

Construction staking is required for any type of construction. Stakes in the ground are necessary to guide the precision building process.

Construction staking is an art form. Some surveyors love to do construction staking and others try to avoid it.

Construction staking can be very rewarding and stressful at the same time.

Most all construction projects are on a tight timeline therefore it can be very demanding of your time. There are huge penalties when construction projects do not meet defined schedules.

COMMUNICATE, COMMUNICATE, COMMUNICATE

It is very important that coordination with the contractor is held on a daily/weekly basis to assist them in meeting their schedule as well as giving you the opportunity to prepare and schedule your field crews in advance to help maintain that schedule. When the contractor or surveyor get off of schedule, it can make for a very stressful project. From my experience, communication between the contractor and surveyor and staying on schedule is the most important part of any construction project.

It is also important that there be one person designated as the point of contact between the contractor and surveyor. I have been on projects where there were multiple supervisors giving orders. It creates confusion and disorganization that slows down the project and ultimately costs more time and money.

At the beginning of any construction project, establish who the point of contact will be, establish a schedule and then stick to it.

APPROVED DESIGN PLANS

Only use approved design plans. You will be taking a big risk performing any staking with unapproved design plans.

AS-BUILTS

As-builts is the final process for most all construction projects. Public agencies will require that the final construction be documented on a set of as-built plans that are certified by a Land Surveyor and/or Engineer.

You will need to coordinate with the contractor and designer to determine what construction elements will need to be as-built with final horizontal and vertical positions. The final measured values will be added to a clean set of design plans next to the corresponding design grades. A cloud is drawn around the added values and are preceded with the letters AB.

MARKINGS ON STAKES

The marking on the stakes as shown herein are representations of only one way of marking them. These markings will need to be adapted to meet the specific needs of the project, contractor and/or surveying company.

For example: Some survey companies will add point numbers, elevations and their companies initials to each lath.

The bottom line is to provide the information that is necessary to build the project. Coordination between the surveyor and the contractor is paramount.

If you have any questions or additions you would like added to future editions of this book, please email me at jcrume@cc4w.net.

ABOUT THE AUTHOR
Jim Crume P.L.S., M.S., CFedS

My land surveying career began several decades ago while attending Albuquerque Technical Vocational Institute in New Mexico and has traversed many states such as Alaska, Arizona, Utah and Wyoming. I am a Professional Land Surveyor in Arizona, Utah and Wyoming. I am an appointed United States Mineral Surveyor and a Bureau of Land Management (BLM) Certified Federal Surveyor. I have many years of computer programming experience related to surveying.

This ebook is dedicated to the many individuals that have helped shape my career. Especially my wife Cindy. She has been my biggest supporter. She has been my instrument person, accountant, advisor and my best friend. Without her, I would not be the professional I am today. Cindy, thank you very much.

Other titles by this author:

http://www.cc4w.net/ebooks.html

Follow us on Facebook

Books available on Amazon.com